Composting
AN EASY HOUSEHOLD GUIDE

Composting

AN EASY HOUSEHOLD GUIDE

Nicky Scott

Green Books

First published in 2005 by
Green Books Ltd, Foxhole, Dartington, Totnes, Devon TQ9 6EB
edit@greenbooks.co.uk www.greenbooks.co.uk

Reprinted 2006

Printed by Cambrian Printers, Aberystwyth, Wales, UK

Text & covers made from 100% recycled paper

ISBN 1 903998 78 6

This book is dedicated to the memory of Steve Portsmouth, who was a leading light in the recycling world and a keen composter. Steve took over the running of South Molton Recycle and made it one of the best community-run projects in the UK.

Thanks all who contributed to this book, especially Jon Clift, Amanda Cuthbert and John Elford at Green Books.

DISCLAIMER: The advice in this book is believed to be correct at the time of printing, but the author and publishers accept no liability for actions inspired by this book.

Contents

Introduction 7

1 Why make compost? 10

2 Compost Happens! 12

3 Getting started 14

Where do I put my compost bin? 14
What can't go in? 15
How do I fill my compost bin? 16
How long does it take to make compost? 17

4 How does it work? 18

5 Composting systems 22

For the beginner
    Dalek-type bin 23
    Tumblers 24
    Digesters 25
    Green Johanna Composters 26
    EM Bokashi – Fermentation 27

Contents (continued)

For the more enthusiastic
 Wormeries   29
 Leaf mould   34

For the serious enthusiast
 Making a hot heap   36
 The New Zealand Box   39

6 Using your compost   43

7 Bigger composting systems   48

8 Getting more involved   53
 Community composting, schools, etc   54

9 A-Z guide to composting   58

10 Resources   92

Introduction

Once you start composting, you get hooked! It is a very simple and satisfying process: not only will your dustbin be lighter, but it won't smell so disgusting – and nor will your compost heap, if you follow the simple instructions in this book.

Even if you do not have a garden, you may have access to a communal green space, which you can either manage yourself or in partnership with others, or your local council manages; and green spaces need compost! I've visited housing estates in London where green spaces are not much in evidence, and there's plenty of composting going on, with the finished compost being used in window boxes, hanging baskets and on open community spaces (see Chapter 8).

You can still make compost in a very small space – even if you only have a tiny garden, or just a balcony or window sill.

Local authorities vary in the amount of help they offer. Many now offer cut-price composting bins, information and even training, supported by compost officers, compost 'hot lines' and regular get-togethers so that people can share experiences and get further help. Why not find out what your local authority offers?

In the past, composting was seen as an activity that only 'gardeners' became involved with, and was surrounded by a huge amount of mystique; many gardeners swore by their secret systems. Nowadays, even people without gardens are composting, because they want the compost for their hanging baskets, window boxes and planters, or because they are keen to reduce waste going to landfill, or both.

Composting has also become a hot political issue. Under European law, Britain must reduce the amount of 'waste' going to landfill, particularly waste which can be composted. To do this, we have to massively increase our recycling and composting. Reducing and disposing of our rubbish can be done in many ways, with incineration and land-filling as the worst options, and home composting as the very best option, since it reduces the amount of material that has to be disposed of in other ways.

You can make a difference. With the enormous problems that face the world, it may seem that our individual efforts are going to be completely insignificant, but many problems we confront are created by individual actions, and we each have enormous powers to make a change. With composting, we can effect change in several ways:

Not only can we stop buying peat-based compost, which causes the destruction of our peat bogs, but we can also make peat-free compost ourselves from materials we commonly put in the dustbin. Composting also helps prevent global warming by reducing emissions of methane, a powerful greenhouse gas which gets produced when organic waste decomposes when buried in landfill sites.

It is estimated that about one quarter of all the UK's methane emissions are due to organic waste rotting in landfill sites.

I'm currently the Chairman of the Community Composting Network. We have as our mission statement, 'Everyone composting!' We feel that just about sums it all up. If everybody were composting, then the environment we all live in, from the local level right up to the whole planet, would be in a far better state. This book will give you all the information you need for successful composting.

Nicky Scott

Why make compost?

There are lots of good reasons to make compost:

- A third to two-thirds of the average dustbin contents can be composted, so you can make a big difference.

- Making compost lightens your dustbin and stops it smelling.

- This means that less is sent to landfill, which in turn means that less noxious liquids and greenhouse gases – especially methane – are produced.

- Composting saves you money – no more peat-based compost will need to be bought!

- Compost improves all soils – it opens up clay soils, and holds moisture and nutrients in light and sandy soils, reducing the need to water and fertilise.

- Compost helps make healthy soils, and healthy soils lead to healthy plants (and people and animals), which are less prone to pest and disease attack.

- Lastly – it's fun, easy and very satisfying!

Compost Happens!

If you still need convincing how simple and quick making compost can be, try this simple demonstration. Be ready to be amazed!

1. Save up some cardboard and paper: toilet roll centres, ripped-up cardboard boxes, cereal boxes, scrunched-up envelopes (minus the windows), kitchen towel – all those bits and pieces which are difficult to recycle.

2. Mow the lawn.

3. Mix or layer grass cuttings and cardboard/paper together thoroughly. Roughly half and half by volume; if in doubt, err on the side of putting in more cardboard/scrunched paper. If layering, aim to get the layers roughly about 25mm (1") thick, but don't be too fussy – just don't put the grass on too thickly!

4. Put in a pile or in a compost container; this can be anything which will contain the heap – it could be a plastic compost bin bought from the council, or it could be a home-made bin made from pallets or scrap wood etc. Repeat every time you mow the lawn.

5. Sit back and wait for compost to happen!

It's as easy as that!

Getting started

I'VE GOT MY COMPOST BIN – WHERE SHOULD I PUT IT?

- Put it anywhere that is convenient, but easily accessible from the kitchen.

- Putting the bin in the sun will speed up the composting process, but it will still work in the shade.

- A compost heap is best sited on soil, but will work on concrete, providing there is some drainage.

- Worms will find their way across tarmac and concrete to colonise your heap! They will like it even more if you have a good bedding layer for them to colonise: moist, scrunched-up cardboard is ideal to put in at the bottom.

- However, if at all possible, place your bin directly on the soil: it's much easier, and encourages the soil micro-organisms to get to work.

WHAT DO I PUT IN IT?

'BROWNS'

- Bits of cardboard, scrunched up paper, loo roll centres, kitchen towel, dried leaves, small woody prunings from the garden etc.

'GREENS'

- Fresh uncooked fruit and vegetable peelings, garden waste (grass clippings, green leaves, soft prunings, etc.)

WHAT CAN'T GO IN MY BIN?

(See A-Z section for what to do with these)

- Glass, tins, plastic – these will never compost!

- Cooked food, meat and fish, cheese fats and grease (they can attract unwanted visitors).

- Roots of any perennial weeds such as ground elder, couch grass, dock and dandelions.

- Any diseased plants

- Cat or dog faeces

All the above can be composted – apart from glass, tins and plastic – but they require experience and care – see the A-Z section.

HOW DO I FILL MY BIN?

FOR QUICK COMPOSTING

- Make alternate layers of greens and browns, each layer being roughly 100mm (4") thick.

THE LAZY WAY

- Just put your greens and browns in the bin whenever you have them – it will just take a bit longer.

NOW WHAT?

All you need to do now is to keep the contents in the bin moist. The best time to check is just before you add more materials.

- If it's dry, water it.

- If it's a bit soggy or slimy, then mix in some more 'browns' such as scrunched-up egg boxes or some torn-up cardboard boxes.

Apart from that, just sit back and relax.

HOW LONG DOES IT TAKE TO MAKE COMPOST?

QUICK COMPOSTING

If you have layered the browns and greens as suggested above, kept it moist and maybe stirred the contents of the bin occasionally, then you should have compost in 3 – 6 months, depending upon the time of year. It's much quicker in the summer.

EVEN QUICKER COMPOSTING

To really boost your composting, chop everything up before you place it in the bin. This also reduces its bulk, so you can also get more into the bin. A shredder is useful for chopping up any woody prunings; they make a great addition to your 'browns' layer.

SLOWER COMPOSTING

Providing you keep the materials in the bin moist and do nothing else, you should have your compost ready to use in 6–12 months. Even if you forget to keep it moist, it will still compost, but will just take a bit longer, depending upon what you placed in it. Compost happens even if you do nothing.

4

How does it work?

EVERYTHING THAT LIVED RECENTLY COULD BE COMPOSTED: Natural products that were once part of a living organism, such as wood and wool, are not only food for a host of insects and other creatures but will slowly rot if the conditions are right. Composting is a way of accelerating this process by creating the ideal conditions for a variety of small organisms, most of which are only visible through a microscope, to do their work.

GETTING THE MIX RIGHT

Composting depends on bringing together the materials that you want to compost, in roughly the right proportions. There are basically two types:

- 'Greens': the wet, soft, green materials (high in nitrogen)

- 'Browns': dry, harder, absorbent materials (high in carbon)

BREAD AND CHEESE

A good analogy comes from the Centre for Alternative Technology: they refer to bread (dry, hard, 'brown', carbon-rich materials) and cheese (soft, wet, 'green', nitrogen-rich materials). This analogy also gives you an idea of the relative volumes that you need: for example, a bucket of kitchen waste needs to be mixed with at least a bucket of scrunched-up paper and cardboard or other 'browns', such as garden shreddings. In fact you can add a lot more paper and cardboard than this – just put it all in your kitchen bin together.

- If you get the proportions roughly right, everything composts readily.

- If it's too wet, it stagnates and goes smelly.

- If it's too dry, it just doesn't do anything much.

- Material for composting needs to be wet, but it also needs plenty of air spaces, much like good soil. If soil is healthy, it will have a good structure with plenty of air spaces, and be damp like a wrung-out sponge. The same conditions are ideal for compost making.

- If you create the right conditions, the bacteria, fungi, and countless other micro- and macro-organisms responsible for composting, will thrive.

You can always add more micro-organisms by adding a light sprinkling of healthy topsoil – or even better, inoculate with some of the previous batch of compost (also see Activators, page 58).

Get the mixture right from the start, and Compost Happens!

Mix dry, absorbent materials into wet sloppy stuff. Otherwise it starts to smell really unpleasant!

Compost needs both air and water – See *Life in the Heap*, page 41.

A pile of dry carbon-rich materials such as branches, sawdust, woodchips etc, is not going to cause a problem by going smelly. It could be left as a long-term heap or a wildlife refuge, or it could be used to mix with green, soft, smelly materials as they arise.

Even if you disregard all the above information, and just dump and run, eventually everything will get eaten by something or other, and converted into a dark crumbly substance that will positively enhance your soil.

The whole secret of compost-making is to set up the ideal environment for the bacteria, fungi and other creatures that are concerned with the decomposition process. They thrive in a moist but not waterlogged environment, with plenty of air: ideally, everything is coated with water but there are air spaces between.

5

Composting systems

There are a bewildering amount of different composting systems and bins available on the market. Why? Are some better than others? Why the need for all this choice? This section will help you choose which bin or system suits your lifestyle, your family, your house or flat.

If you live in a flat, or do not have access to a garden or outside space, you can use a Bokashi system or wormery to deal with all your kitchen waste (see pages 27–34). Finding an outside space to compost might be a problem, but it can be done (see *Community Composting* on page 48 for ideas on small space composting.)

Composting systems, boxes and containers broadly fall into two types:

 The first type deals with fresh, uncooked fruit and vegetable skins and peelings, cardboard and paper, as well as green garden waste materials – prunings, hedge clippings etc.

The second type deals with all but can also be used for other food wastes as well, such as cooked food, meat, fish, cheese, fats and grease.

COMPOSTING SYSTEMS FOR THE BEGINNER

DALEK-TYPE BIN

The compost bin that most people are familiar with is the plastic 'Dalek'-type bin, promoted by many local authorities. Sizes vary from just over 200 litres to over 700 litres; some have access/inspection hatches, and they come in a variety of colours. Millions of these are now in use in the UK.

These bins are available through many councils, water companies and garden centres.

TUMBLERS

Because uncooked fruit and vegetable waste is dense and wet, one way to deal with it is to aerate it by putting it in a tumbler. A tumbler consists of a drum mounted on a stand; they either tumble end over end, or around on their axis. They are also useful for dealing with perennial weeds, and for mixing materials. However, they take up a lot of space. The coarse compost they produce can be used directly on the garden, or can be placed in a covered pile in your garden to enable it to mature to a finer product.

Tumblers are available through some councils and large garden centres.

DIGESTERS

The most common digester is the 'Green Cone'. It consists of a basket, rather like a washing basket, which is buried in the ground with a double skin cone, which is all that is visible above ground. This makes it difficult for rats to get in. The material breaks down and is pulled into the surrounding soil by worms. Since kitchen waste is largely liquid, much of this also goes into the soil where the nearby plants can take it up.

A digester is more of a waste disposal option, since you don't harvest the compost. Digesters have to be installed carefully, and will need moving after a year or two, depending on your soil type and how much you put in them, unless you are very frugal in their use.

Green Cones are available through some councils and some big garden centres: see www.greencone.com.

GREEN JOHANNA COMPOSTERS

Green Johannas are the 'Rolls Royce' of the plastic compost bin. These composters are fully sealed with a base plate so that rats cannot get in. They come with a range of accessories, including a 'duvet' to keep your compost warm in the winter – it really helps to insulate your compost during the cold months.

The great advantage of this large system is that you can put anything in – you don't have to separate out kitchen waste from garden waste, and the manufacturers claim you do not need any composting experience.

Green Johannas are extremely user-friendly. See: www.greenjohanna.se.

EM BOKASHI – FERMENTATION

The Effective Micro-organism 'EM Bokashi' system is an airless system, which uses bacteria that thrive without air to ferment the material. Unlike most airless systems, EM Bokashi does this without unpleasant smells, and can be placed and used within the kitchen or some other warm part of your house or flat. (Many foods are preserved using fermentation processes, such as yoghurt, sourdough bread, beer and sauerkraut.)

The Bokashi system is well proven and used extremely successfully in many countries. For example, in Korea alone over 3 million households use this system to deal with their kitchen waste.

This is how it works:

- There are two pairs of buckets. Each pair fits together – the one on top has holes drilled in the bottom, so that liquids can collect in the second bucket underneath.

- You will also get a bag of bokashi mix with the kit: a combination of bran and friendly micro-organisms.

- Every time you add material (e.g. food scraps) to the top bucket, add a little sprinkling of the bokashi mix, push the material down firmly to squeeze out any liquids and remove as much air as possible, and re-seal the lid.

- When the top bucket is full, start using the second set of buckets

- When the second set is full, empty the first set and use that.

- You can either bury the contents, which very rapidly break down in the soil, making a great organic fertiliser for your plants, or add them to your compost heap, where they are no longer attractive to rats (see page 51).

- The liquid that gathers in the bottom bucket can also be diluted and used as a wonderful organic liquid fertiliser for your house and garden plants.

- You will need to buy the bokashi mix every so often.

For more information on EM Bokashi, see www.livingsoil.co.uk, www.effectivemicro-organisms.co.uk and www.emtechnologynetwork.org.

COMPOSTING SYSTEMS FOR THE ENTHUSIAST

WORMERIES

This system requires a bit more effort, but is great fun and children love it!

Worms eat rotting matter, and are particularly useful because they will eat your food waste, paper and cardboard. Their manure, called 'worm casts', is very beneficial for all soils and plants. It is used more as a fertiliser than as a bulky soil improver.

Wormeries also produce an extremely valuable liquid fertiliser, which is drained off at regular intervals. When diluted with water (at least 10 parts water to one part liquid), it is an excellent feed for flowers and vegetables. If you don't regularly tap off the liquid, the container will gradually fill up with it and drown all your worms – and knock you out with the odour!

Worms like it to be cool and moist, but not too cold. They will not be very active at low temperatures, and if it gets too hot, they will climb out if they can. If they get too wet, they may drown or migrate.

Buying your wormery

Starter Kits If you know nothing about wormeries, consider buying your wormery complete with a starter kit – this will contain all you need, including worms and instructions, to get you up and running.

Wormeries The majority of wormeries are made out of plastic, and there are many to choose from – from single containers to stacking systems which are designed to make it easier for you to extract the worm casts. Whatever type you buy, make sure there is a tap at the base to drain off excess fluid.

Your local council may be offering a specific type at a reduced price, or you can easily research types and sources from one of the many sites available on the internet. For starters, try:

www.sunshineways.com/worm.html
www.greengardener.co.uk/worms.htm
www.westcountryworms.co.uk/
www.thewormhotel.com/wormery.htm
www.originalorganics.co.uk/wormeries.htm

You can also buy wooden wormeries. See:
www.woodsfarmorganics.co.uk/worms/wastebusters.php

Making your own wormery

A container When you are starting from scratch, as with any animal, you have to provide a suitable living environment.

If you want to make your own wormery, many large containers (such as old dustbins and barrels) can be adapted to become wormeries.

For ideas on making your own wormery see:
www.troubleatmill.com/wormbin.htm

Bedding No matter which container you use, you must start the worms off with a generous bedding layer. Bedding can be leaf mould, finished compost (preferably sieved), shredded-up newspaper and/or cardboard, well rotted sawdust or woodchip, or a mixture of any or all of these. Whatever it is, it must be thoroughly wetted – especially paper and cardboard, as worms will die if they dry out.

Getting your Worms Now you've got somewhere for them to live, all you need is your worms! Don't be tempted to dig worms out of the soil in your garden for your wormery; they will not be the right ones. You need compost worms, such as tiger worms or dendras, which live naturally in compost and manure heaps – but you need an awful lot of them.

The best way is to buy in enough worms to really kick-start your wormery into action – at least 500 but preferably 1000 worms should do the job. Most of the companies that supply wormeries will also sell you worms. They arrive in a container through the post, ready for action. See any of the suppliers above or:

www.wormsdirectuk.co.uk/
www.wigglywigglers.co.uk/shop/

Feeding your worms

- After you have introduced the worms into the container, let them settle down for a day or two. They will be quite happy eating their bedding.

- Only feed them small amounts at a time: they don't want a great pile of stuff dumped on them, as it can compost and generate heat – and they like it cool! Little and often is best.

- Worms can eat about their own weight in food each day. The more worms you have, the faster it all happens.

Be patient – your wormery will probably take at least a year to get up to full speed as the worms breed. You can expect 15,000 – 20,000 worms in your wormery by then!

What to put in your wormery

Wormeries are ideal for small amounts of 'difficult' kitchen materials – food scraps, cooked leftovers, meat and fish, cheese rinds, bread. Avoid large amounts of raw fresh fruit, vegetable trimmings, and garden waste (ideally these will go in your garden composting system, if you have one).

If you separate out your materials so that fresh vegetable trimmings mostly go in your compost heap and the kitchen scraps go in the wormery, then you won't create masses of liquid.

Harvesting – collecting the worm compost

Wormeries take a long time to fill up with worm casts. When they are getting pretty full, remove the freshest material plus the layer immediately underneath – this will contain most of the worms. Put this to one side. It can all go back into your wormery when you have harvested the worm casts – the rich dark material at the bottom.

If you have a stacking system, such as the popular 'Can-O-Worms', you merely remove and empty the bottom container, and place it back on the top of the wormery. The whole cycle then starts again. See: www.abundantearth.com/store/canoworms.html.

Using worm casts

Worm casts (the name given to the finished worm compost) are the *crème de la crème* of composts, and are best used by the handful rather than the wheelbarrow load. Think of worm casts as fertiliser, not compost: a little goes a long way. Give all your pot plants, window boxes, hanging baskets a top dressing. Water them thoroughly first, and then top dress with a handful or so of worm casts. You can do the same with garden plants.

Using the liquid fertiliser

The liquid that you drain off from your wormery makes a wonderful liquid feed for all your plants, especially fruit – dilute with about ten parts water before using as a foliar feed or to water your plants with.

LEAF MOULD (COMPOSTING LEAVES)

Leaves can be composted in small quantities in any composting system (see pages 23–39).

FAST LEAF MOULD (COMPOSTING LEAVES)

You can speed up leaf mould production by picking the leaves off grassy areas using a lawnmower. This chops up the leaves and mixes them with grass, which is high in nitrogen. The grass is also mostly water, so you end up with the magic mixture of carbon and air (in the leaves), nitrogen and water (in the grass cuttings), and micro-organisms (on both). If you make a decent pile of these, they will heat up!

CARDBOARD MOULD

You can also layer cardboard with grass cuttings to make a wonderful leaf mould/peat substitute. Just make a layered stack of thin layers of grass with flat cardboard sheets between.

PEAT SUBSTITUTE

Whatever method you use, you will end up with leaf mould. This resembles peat, and can be used in much the same way, but unlike peat it is a renewable resource. Leaf mould is an invaluable ingredient in seed and potting mixes (see *Using Your Compost*, page 43), so don't let those leaves go to waste! If you only have very small amounts of leaves, they can just be added to any of the composting or fermentation systems.

COMPOSTING SYSTEMS FOR THE SERIOUS ENTHUSIAST

As your confidence and understanding of composting increases, you will inevitably want to increase the range and amount of materials you compost. Whilst 'nearly everything that once lived could be composted', certain materials present us with challenges, and just about anything in large quantities can be a challenge. Once you feel more confident you can move away from making compost in a plastic container and make your own compost heap, or even a 'hot heap'.

THE HOT HEAP

This system gives you the opportunity to make much larger quantities of excellent compost quickly, to use on your garden or allotment. It's also fascinating and fun.

When all the conditions are right, a compost heap can get really hot, cooking weed seeds and roots, and rapidly killing any potential diseases or pathogens. However, the heap needs to be big enough or very well insulated, as if it isn't this heat will rapidly dissipate.

Making a hot heap

To make a hot heap, first gather all your materials together – the best time is when the growing season is in full swing, when you can go and gather plant material to bulk up your heap.

BECOME A COMPOST FORAGER!

- Nettles and other weeds can always be found somewhere – neighbours might be happy to help you with materials
- Perhaps you have bags of leaves saved from the autumn, or a nice pile of partly rotten wood chip
- You might be able to get hold of some strawy manure
- Your local hairdresser will happily give you hair
- The greengrocer will give spoiled fruit and vegetables
- The brewery will give you hop waste, and so on
- Choose a time when you can add hedge prunings and grass cuttings
- Don't forget to add cardboard and paper.

A shredder can be really useful for the tough and bigger stuff!

Build your heap

Option 1: No box – a pile on the ground

Assemble as much material as you can – ideally enough to make a 4' x 4' ((1.2 x 1.2 metres) roughly cube-shaped pile. You can heap the materials up as high as you can reach – it will end up being conical, or if you are really neat, a pyramidal shape. If you have more materials

than this, then extend the heap sideways: this then forms what is known as a windrow, and is the way to compost really large quantities.

Option 2: Put it in a box

You can make a cheap simple box to contain your heap out of old pallets. These can simply be tied together, and you can easily insulate them if desired: fill the space where the pallet is picked up by the forklift with rolled-up cardboard, carpet, bubble wrap, an old quilt, sheep's fleece, an old coat or suitable building insulation material, and line the inside with cardboard sheets to cover up the gaps between the slats of wood.

Option 3: Use two boxes

The 'Rolls-Royce' design for this type of heap is the New Zealand box. You can buy one ready-made, or construct your own.

Fill up your boxes

- Start with a layer of material high in carbon (i.e. the tougher, drier stuff): not too thick – a layer of 75–100 mm (3–4").

- Follow with a thinner layer that is high in nitrogen, e.g. grass cuttings. Denser green materials such as kitchen waste or grass cuttings should be layered thinly, approx. 25–50 mm (1"–2"), whilst looser, more open green items such as tomato stems or cabbage leaves can be applied in thicker layers, about 50–100 mm (2"–4").

- If the materials are on the dry side, add water as you go.

- Adding small amounts from a previous heap will 'inoculate' the heap with countless micro-organisms. So if you have it, mix in the uncomposted top layer of a previous heap.

- Repeat the layers as high as you can go – even beyond the height of the sides. It will soon sink down.

When the heap is built, cover it with some old plastic sacks or sheeting to keep in the water vapour that will be given off, and some old carpet that will help to keep the heat in.

Monitoring and turning

A large heap made in one go will rapidly heat up in a few days, and then start to cool down. Keen composters have a thermometer on a probe, so that they can monitor when the temperature drops. At this point, they turn the heap.

The advantage of doing this is that you can mix all the ingredients: you can move the materials that were on the outside towards the middle, and vice versa. (Do not turn large, musty, dusty heaps. Wet them down first, or wear a dust mask to avoid breathing in the fungal spores.)

You can also check the moisture content, and adjust if necessary. After turning, the heap should re-heat, and then it can be turned again, to accelerate composting.

Troubleshooting

After a day or so, the heap should be very hot: when you take the covers off, you should be able to see the steam rising. If not, the mixture is not right:

- It could be too dense and wet. If wet, it needs opening up. Re-mix, adding something to absorb the water to let the air in and to add carbon.

- It could be too light and airy: If dry, it needs watering, or remixing with something green and sappy.

LIFE IN THE COMPOST HEAP

What is astonishing is the amount of life that a compost heap supports.

A gram of healthy soil (roughly a teaspoon) contains about a billion microscopic organisms: predominantly bacteria and fungi, but literally thousands of different species. In the top nine inches of soil there are other larger fauna – mites, springtails, spiders, ants, beetles, centipedes, millipedes, slugs and snails – 10,000 to 100,000 per square metre, and that's just in the top 9 inches! And of course there are earthworms (30–300 per sq. metre) – these are not the same as the worms found in compost heaps. Now that's a staggering amount of soil life in each gram, in each square metre, and in each acre.

And compost heaps are richer than soils. Given the right conditions, colonies of bacteria in a compost heap can double every hour – and keep on doubling, reaching astronomical numbers. It is primarily this phenomenal rise in the numbers of bacteria, all respiring, consuming and reproducing, converting the energy stored in the materials in the heap, which gives rise to the heat.

Although the thermophilic (heat-loving) bacteria are the ones that really thrive in the heat, it's not just bacteria that multiply in the heap. As the heap cools, there's plenty of food to go around for everyone, and compost heaps become a magnet for all kinds of creatures, including some bigger ones: predatory beetles move in on larvae and smaller creatures; frogs and toads do likewise, as do slow worms and grass snakes, which love the warmth in a heap. Birds will visit to pick off insects and larvae, and bats will even visit at night. For wildlife value alone, it's worth making compost.

6

Using your compost

Finished compost is wonderful stuff, and you can never have too much of it. Compost from your garden system is a wonderful soil improver, adding water-holding capacity to light, free-draining soils, and helping to break up heavy clay soils. It is likely to have the odd weed seed in it, so if you use your compost as a top dressing remember to go around and hoe the weeds as they germinate.

IN CONTAINER PLANTS

Use compost as a top dressing for all your container plants (inside and out), hanging baskets, window boxes, etc. Thoroughly water the plants first, and then apply the compost. If the plants are looking a bit tired, move them up a size of pot; this will enable you to add a good amount of fresh compost when re-potting.

Wormery compost is much richer than compost from your garden bin, and is used by the handful, rather than the barrow load. It's great for pepping up your container grown plants – see below.

IN THE GARDEN

Use compost in the garden; apply liberally to all beds, giving a good top dressing. Water first, or ideally apply after a good amount of rain. Dig in lots of compost when you are planting new plants.

IN THE SOIL

Dig compost into the soil to improve the structure, or just put it in where you are going to sow or plant.

AS GROWING MEDIA

This is the stuff you buy at a garden centre in a bag labelled 'compost'.

All kinds of materials have been used to make commercial growing media, which is the term given for seed composts, hanging basket compost, ericaceous compost and so on. Many of these mixes have no actual compost in them, being composed mainly of peat and chemicals.

Peat is an irreplaceable precious resource, which can easily be substituted with materials that are often wasted.

MAKING YOUR OWN GROWING MEDIA

Common materials that can be used to make your own growing media mix include:

- Leaf mould – leaves from trees left to rot down either in a contained pile or in black plastic bags for a year or so. See Leaves & Leaf Mould, page 78.

- Sharp sand and gravel

- Sieved compost

- Molehills or good sieved soil

- Vermiculite

- Perlite

- Bark chippings

Help preserve our peat bogs – use your compost to make your own growing media.

See www.ipcc.ie for more on peatbog conservation.

MAKE YOUR OWN SEED COMPOST

You can make your own seed compost by composting grass cuttings with cardboard to make a very creditable peat substitute (see *Leaf Mould Cardboard*, page 35). If you use soil or compost in a seed mix, you really need to sterilise the soil and be sure that your compost is made really well with lots of heat, as in a hot heap, so that any weed seeds are dead.

MAKING YOUR OWN PLANT COMPOST

You can probably just use your own sieved compost for growing on plants – this is a matter for some judgement on your part. Use your senses: feel the mix, squeeze it, smell it, put some in a pot and water it, and see what it does. If it does not drain easily, then mix in some sharp sand or gravel – remember to always put some drainage in the bottom of the pot, especially if you are using terracotta pots. If it drains too easily, then add some good molehill soil or good sieved soil or/and leaf mould. You will undoubtedly get some weeds popping up, but these are easily removed.

MAKING YOUR OWN CUTTINGS COMPOST

For cuttings, you need to 'open up' the mix with plenty of sharp sand: mix the sand half and half with leaf mould or sieved compost, and stand the pot in a saucer of water to keep moist. If you put your cuttings in, then put two or three sticks around the edge of the pot and cover the whole with a clear plastic bag, you will have a mini-propagator.

7

Bigger systems for several households or communities

COMMUNITY COMPOSTING

This is where a group of people, maybe from the same street, cul-de-sac or even village get together to compost their 'waste' materials. This sociable and enjoyable activity benefits the whole community, turning what were unwanted materials into a valuable end-product. Sharing labour and tools as well as organic waste enables successful composting, and is far more likely to attract funding, should you need it.

Community composting projects range in scale from small groups working on allotment sites or promoting home composting, to social enterprises with local authority contracts providing kerbside collection services. The common theme is that the local community is involved in the management of the organic waste it produces, and its organisations are not-for-profit and locally accountable. Many projects work with social services, helping provide therapeutic employment for some of the most disadvantaged people in the community.

Why not join or form a community composting group? Find out more by contacting the Community Composting Network (see page 92 or visit www.communitycompost.org/aboutus/index.htm).

COMPOSTING WITHOUT GARDENS

Many of the people helped by the community composting sector are people without gardens.

There is one system of composting which is particularly suited to large numbers of people living in cities, including blocks of flats: in-vessel composting. This system is completely sealed from flies, rats etc. It has an air flow system and often a mechanised system for moving or turning the contents, again to assist in aeration. Some systems also have heating to kick-start the composting process (see Southwark's community project below).

SOUTHWARK'S COMMUNITY PROJECT

On a housing estate in Southwark, London, there is a community project which has a small communal garden over the road, but the estate itself is surrounded by tarmac and concrete.

The composting unit is housed in a smart stainless steel box which contains two composting sections. A grid on the bottom allows any liquid to drain off into a pipe, which in this case leads to the sewer.

When the box is opened, it's easy to see the finished black compost at the bottom, and a layer thick with worms just under the fresh food layer. The worms manage to find their own way into the composting unit, which is in the middle of an area of tarmac and concrete! This system is really a sealed 'hybrid' system. The top section is hot composting, but because relatively small amounts are being added all the time and it isn't an insulated container, the environment is perfect for worms, which can get away from the heat by burrowing down, if they need to.

The local council helped the residents to set it all up, and provided them with window boxes and hanging baskets, as well as planning the communal garden over the road.

HACKNEY'S BOKASHI SYSTEM

The E M Bokashi System is being used extremely successfully on a high-rise estate in Hackney, London. Residents on the estate had problems with the garbage chutes blocking up, causing food waste to accumulate, which created horrible smells that attracted flies, rats and foxes.

A community composting project on the estate offered residents a collection service. They gave all participants a set of buckets and a supply of Bokashi mix. In addition, they collected the cardboard, which was the main reason why the chutes had become blocked.

The residents now use the Bokashi system in their flats to collect and treat their kitchen waste. The fermented kitchen waste, together with the cardboard, is collected weekly. The cardboard is shredded and mixed with the kitchen waste before being loaded into an 'in-vessel' composting machine called the 'Webbs Rocket'. This composts the material for two weeks before it is put into large-scale wormery systems, which breaks it down even further so it is ready for use.

This project has been so successful that the council have asked the community composting organisers to expand their service to more estates, and have recruited thirty-five new staff members. For more information on this project, see:

www.elcrp-recycling.com/bokashi.html
www.quickcompost.co.uk.

Several councils now use huge industrial versions of the systems used in these projects to process and compost thousands of tons of green waste, which they collect from households through their kerbside collection schemes. The finished compost is then either bagged up and sold for domestic use or collected in bulk and used in horticulture and landscaping schemes.

You really don't have to have a garden to make compost! If you live on an estate, maybe your local council could help you out too – contact your local recycling officer and tell him or her what is being done in other places, and ask what can be done for your estate. You can also contact the Community Composting Network for help and advice.

Even if you can't get any help from elsewhere, you could still have a wormery if you have a balcony or any outside space – that way you can make some compost to grow a few plants in.

8

Getting more involved

Composting is such a satisfying and therapeutic activity that many people become positively evangelical about it. Once you have seen the light, it's difficult to see anything that could be composted being consigned to the dustbin and landfill.

Nothing can give you the confidence about how to make compost as well as another person. There is no substitute for seeing how it's done, getting your hands dirty and doing it with other enthusiastic composters, sharing their knowledge.

All sorts of community groups, councils and organisations welcome people who want to get involved in spreading good composting practice. Becoming a volunteer with a community group can often lead to paid employment; your local council may well pay to have you trained to work with projects in schools, or as a master composter in conjunction with WRAP (see *Resources*). Or you might want to start your own project.

Many community composting projects are mental health or learning disabled projects which provide enjoyment for all involved.

See www.thrive.org.uk, or contact the Community Composting Network www.communitycompost.org.

SET UP A COMMUNITY HOME COMPOSTING PROJECT

If you feel you would like to do more, perhaps by setting up a community composting project, contact the Community Composting Network (CCN) – see Chapter 7, *Bigger systems*, and the *Resources* section. Some community composting projects concentrate on home composting, demonstration and education. CCN can help to guide you to your nearest community project or help you set up your own project.

In London there is a 'master composter' campaign run by the London Community Recycling Network: email them at compost@lcrn.org.uk or write to London CRN Compost Networks, The Grayston Centre, 28 Charles Square, London N1 6HT. Also see: www.lcrn.org.uk/programmes/compost_networks/ composting.

SURREY: Because of problems finding a suitable site for composting, one project in Surrey now operates out of a back garden, hosting regular demonstration sessions. They have also cleverly adapted a trailer as a mobile display unit and go to all kinds of events promoting the wonders of composting. See www.compostworks.org.uk/

LANCASHIRE County Council has had a major campaign employing a community composting member to help distribute 60,000 composting bins (for free!) over three years, backed up by a composting 'hotline' and a whole range of information and support. See www.compost-it.org.uk/

BECOME A MASTER COMPOSTER

The Government has become aware of the increasing importance of composting and recycling, and has set up an organisation called WRAP (Waste and Resources Action Programme – see *Resources* section for contact details). WRAP now has a home composting campaign, which works by forming partnerships with local authorities who then train local people to become 'master composters'. Trained master composters can help promote home composting very effectively.

Find out more by contacting your local council or by looking at the WRAP website: www.wrap.org.uk.

GO BACK TO SCHOOL

Young kids love the 'yuk' factor of composting, and all the associated creepy-crawlies. If you can get the kids really clued up about composting, then 'pester power' will hopefully make its way to changing habits at home.

Some councils are involved with going into schools to promote recycling generally, so if you are a teacher, or want to get involved with composting in schools, contacting your local council is a good first move.

There are groups who promote healthy eating in schools, who sometimes liaise with the CCN to show children how to compost their leftovers – apple cores, banana skins, etc.

See www.wen.org.uk/local_food/schools.htm – this website shows you how to get a schools scheme up and running, offers useful links to the National Curriculum, and shows how composting can be incorporated into key studies. Also see www.hdra.org.uk/schools_organic_network/

DEVON's schools were invited to design web pages or make videos showing how easy it is to turn kitchen scraps and garden waste into top quality compost for the garden. There was a fantastic response to the composting competition, and judges found it hard to choose a winner.

The competition was organised by Devon County Council and the Environment Agency on behalf of the Devon Organic Waste Working Group. The Environment Agency sponsored the top prize of £1,000 and the County Council sponsored second (£200) and third (£100) prizes. See www.recycledevon.org/pages/info_composting.asp.

Why not find out whether your council provides an opportunity for your school to get involved with composting?

The WILTSHIRE Wildlife Trust has an exciting project involving in-vessel composters called Rockets (see page 51), which safely compost food waste generated from school dinners and packed lunches. They also have a team of Compost Ambassadors who can help with all Wiltshire's composting needs. See www.wiltshirewildlife.org.

BRITAIN IN BLOOM

Communities all over the UK get involved in Britain in Bloom, and thousands of containers, hanging baskets, troughs and window boxes get filled with compost. The Britain in Bloom rules now have recycling and sustainability criteria, so the judges are looking to see if the group is using compost it has made itself, as well as its general awareness of recycling and wildlife issues. They award extra points if the compost is home-made. See www.rhs.org.uk/britaininbloom/sustainability.asp.

9

A-Z guide to composting

Activators

An activator is something added to a compost heap to speed up the composting process – however, compost made correctly doesn't generally need an activator. Commercially available activators are generally either chemical nitrogen, which can have a detrimental effect on the micro-organisms in the heap, or a bacterial culture.

If you want to introduce more bacteria, the easiest way is to add some of your old mature compost, if you have some, or add the odd light sprinkling of soil, which will introduce countless millions of bacteria.

Nevertheless, for those who want to use activators:

· Urine is probably the cheapest and best (dilute 1:4).

· Grass cuttings in thin layers are excellent.

· Seaweed – either freshly harvested, as liquid concentrate, or as a dried meal.

- Manures – layer or mix into the heap. Remember that manures are pretty strong, especially chicken manure, so always try and break them up, and add in thin layers.

- Biodynamic gardeners and farmers, who follow the teachings of Rudolph Steiner, put special herbal preparations into their heap. These can be bought from the Biodynamic Association (see *Resources* section).

- Maye E. Bruce formulated a simplified version of this, consisting of the main herbs used in the biodynamic preparations in the 1930s. This can still be obtained through the Henry Doubleday Research Association, and is called QR, which stands for 'quick return'. The main herbs in QR are camomile, dandelion, valerian, yarrow, nettle, and powdered oak bark. Be careful if you add these as live plants, since some are quite pernicious weeds, and small bits of root not properly composted could take hold and thrive in your compost.

- Nettles compost extremely well, and are a valuable addition to any compost heap, but don't add the roots.

- You can also buy various proprietary activators for individual materials, e.g. for grass cuttings, leaf mould etc.

Additives

– also see Weeds.

Various materials can be added to composting and wormery systems to supply missing nutrients or to balance the acidity/ alkalinity:

- **Limestone**, which counteracts acidity, must never be sprinkled directly onto manures, as you will get a release of ammonia and nitrogen as a gas. Compost heaps and wormeries that are too wet and airless become smelly and acid – by adding more dry material and some limestone you will correct the balance.

- **Rock dusts**, e.g. dolomite, which contains calcium and magnesium and is useful to correct over-acidity in wormeries and other kitchen waste systems.

- **Wood ash** which is a rich alkali source – but don't add anything to a heap in great dollops or it will turn to a horrible sludge; instead scatter it and mix in well.

- Crushed up **bones** and **eggshells** (see Bones and Shells).

Air

For composting to work, there must be sufficient air for the organisms that are living in the compost heap to thrive. The air is primarily contained in the 'Brown' materials you add to the heap and the air spaces they create as they decompose (see page 63).

Anaerobic

This means 'absence of air' and is what happens, for instance, when you make a pile of grass cuttings. The mass heats up and starts to collapse and pack all the air spaces together, until only the smelly 'anaerobic' bacteria can survive – to give composting a bad name. There are, however, good anaerobic micro-organisms (see Bokashi).

Ants

A few ants in the compost heap are not doing any harm – in fact they are probably helping it. If you want to harvest the compost, give it a good watering the day before, and the ants will move out. However, if you live near woodland and your heap is invaded by wood ants, they will probably steal your entire heap, and if you put compost out as a mulch around your plants, they will steal all that too!

Ash

See Coal Ash and Wood Ash.

Autumn Leaves

Leaves are so easy to compost and so good for the garden that you really shouldn't burn them. To find out how to deal with them, see Leaf Mould.

Bindweed

See Pernicious Weeds.

Bokashi

This is an anaerobic or fermentation system of composting. EM Bokashi (the EM stands for effective micro-organisms, and is a mixture of different types of bacteria and fungi) was developed in Japan. See page 27 for more on Bokashi.

Bones

Bonemeal is a natural source of phosphorus for the gardener. Some people say that the origin of the word bonfire is a fire for bones – they certainly are a lot easier to grind up when thoroughly dried out or heated up.

To make bonemeal:

· Boil up all bones for stock

· Dry out the bones: if you have a cooker that is always on, like an Aga or Rayburn, put them in the bottom oven or put in any low oven – utilise the heat after cooking, for instance.

· Periodically remove them and grind up with a large pestle and mortar.

· Use as an additive: sprinkle into your compost or wormery.

If you don't break them up, they take a very long time to break down!

Bread

You can put bread in any of the kitchen waste composting systems described in this book, but you can also try to reduce how much you waste. For example, turn stale bread into breadcrumbs, which can then be mixed with left-over fat to fill coconut halves and put outside to feed the birds; or use the bread or breadcrumbs in cooking – think croutons, bread sauce, stuffing etc.

'Browns'

See Carbon below, and *How do I fill my bin*, page 16.

Cans

Cannot be composted; if you do not have a kerbside collection, recycle them at your local recycling centre.

Carbon

Carbon is the main building block of life. In composting terms its most useful common form is as cellulose, such as paper, cardboard and the fibrous parts of plants.

Materials high in carbon ('Browns') do not compost readily without being mixed with materials high in nitrogen – or 'Greens'. This means that you can stockpile them safely to be used when needed, unlike the 'green' materials, which really need to be dealt with straight away.

Carbon to Nitrogen Ratio (C:N)

To make the best compost, you need to have the right carbon to nitrogen ratio in your heap: the optimum ratio is 25/30:1. The easiest way to understand the correct balance between 'browns' and 'greens', and therefore to get the right ratio, is to look at what the C:N ratios are in some common materials:

- Vegetable waste and grass cuttings are about 12:1 i.e. there are 12 parts carbon to each part of nitrogen – over twice as nitrogen rich as the optimum ratio.

- At the other end of the scale, fresh sawdust is around 500:1.

- The pods from peas and beans, at 30:1, are about perfect to compost on their own – plenty of fibre, and yet still green.

- In practice, just think of one bucket of kitchen waste or grass cuttings to at least two buckets of scrunched-up paper or ripped-up cardboard.

Cardboard – Natural and Coloured

Large sheets of cardboard are really useful if you have a garden and a large compost heap.

- Any type of card or cardboard will compost – just make sure that it is mixed with sufficient 'Green' material and kept moist. It can go in any composting or wormery system.

- You can even make a heap just of cardboard sheets and grass cuttings – an inch or two of grass, then a sheet of cardboard. Carry on sandwiching the two materials – you don't even have to rip up the cardboard!

- Line compost heaps with it – especially home-made bins made from old wooden pallets. It stops material going between the slats, and helps to insulate the heap.

- Sheets can be put down around perennial plants as a 'barrier mulch' which suppresses weeds; you can hide the cardboard with a layer of compost or soil etc.

Carpet & Carpet Underfelt

Only pure, natural carpets and carpet underlay can be composted (or successfully used as mulches): wool, silk, jute, hessian etc. A square of carpet on top of a compost heap is useful as insulation and moisture control. Test a few threads of the carpet with a match; if they melt, it isn't natural!

Cartons

- Waxed cartons will eventually compost, but they take a long time. Many cartons and frozen food boxes have a thin skin of plastic, which is a nuisance in the compost, although you can sieve it out.

- Or you can make a separate heap just for this kind of material – mix with grass cuttings.

- Alternatively, they make excellent flower pots: a litre carton will support a tomato plant and you can even plant the whole thing out. Don't forget to cut drainage holes or remove the base.

Cat Faeces

See under Dog and Cat Faeces.

Cat Litter

Cat litter is compostable only if it is made from a natural product – see Dog and Cat Faeces.

Chicken Manure

Use as a compost activator (see Activators), but always mix manures up with 'brown' material to introduce air and carbon.

Citrus Fruit

See Fruit.

Coal Ash

Although coal was formed from living organisms, it was at a time when our atmosphere was different: coal ash contains high levels of sulphur. Very small amounts will not be a problem, so if you have the odd piece of coal in a wood fire don't worry, but if you are burning coal all the time then it will poison the soil. (Wood and charcoal ash, however, are fine for composting; see Additives and Wood Ash.)

Coffee Grounds

These are a good compost activator – get them from cafés for composting if you can! If you have large quantities, you will have to find other materials to mix with them. Coffee grounds are said to be a good slug and snail deterrent.

Comfrey

Comfrey is a valuable compost activator (being very high in Nitrogen, with a 9:1 C:N ratio) and as an additive, especially the hybrid varieties of Russian Comfrey that Henry Doubleday devoted much of his life studying. It is a vital part of any organic garden.

Comfrey is one of the highest natural sources of potassium, and is best used as a liquid, which you can make yourself:

To make comfrey liquid :

- Chop the whole plant off at the base (it will soon grow again!),

- Push the plant into a barrel which has a tap on the bottom, and force it down with a spade to get as much in as possible.

- Put a weighted lid on top, and catch the liquid that comes off in a bucket.

- Water down the liquid (1 part liquid to 10 parts water) and use as a foliar feed – great for tomatoes and all fruit.

For more information see the Henry Doubleday Research Association (HDRA) website: www.hdra.org.uk.

Compost

This can be:

- The stuff which you make in your bin, heap or wormery by composting your waste.

- Growing media that you buy, which has nothing to do with the composting process and is often no more than 'peat in a bag with added chemicals'.

Composting is incredible in the way that it purifies diseased matter, but you must treat it with respect. Wash your hands carefully after handling compost – especially if you have been more adventurous with what you compost.

Compost and Plant Diseases

Many plant diseases can be suppressed by the application of compost, including clubroot (which affects the cabbage family), white rot (which affects the onion family), brown rot (which affects potatoes), and many field crop diseases.

Compost Tea

Some authorities are now claiming that 'compost teas' are beneficial.

To make compost tea:

- Put a sack of compost in a barrel of water for a few days

- Remove the sack, squeezing most water out

- Strain and dilute 1:10

- Use as a foliar feed on your plants. Don't spray plants in full sun, and for maximum effect add a drop of liquid soap (not detergent) and aim to coat leaves both above and below. NB: it's for the plants – not to replace your own cuppa!

Similarly, a useful 'juice' is the liquid you can tap from a wormery or a Bokashi system.

Contaminants

People are naturally concerned that they are not contaminating their compost with any toxic polluting chemicals. Many of the foods we eat contain (or have been sprayed) with chemical pesticides and preservatives, and paper and cardboard will contain amounts of chlorine, boron, and some chemicals from inks, although most inks nowadays are made from vegetable-based dyes.

The composting process itself is incredibly effective at dismantling complex, man-made, carbon-based chemicals, and different bacteria will eat a whole range of pollutant materials, including diesel and old tyres – this doesn't mean you can chuck old tyres onto your compost heap!

Cooked Food

Do not put cooked food into either a 'Dalek'-type or an open garden bin such as a New Zealand box. However, it can go into several composting systems, such as tumbler compost bins, wormeries, Bokashi systems, digesters and Green Johanna-type bins. See under *Composting Systems*, page 22).

In FLANDERS, the local authority will either supply you with a home composter, or a pair of former battery chickens to eat left-over food scraps. Here in the UK, government legislation says that if you keep chickens, any food waste coming from a non-vegan kitchen must be composted in a sealed container that the chickens cannot gain access to.

Having pets, especially dogs, is a great disposal route for most cooked foods, and if you don't have a dog, donate them to a neighbour's dog!

Cooking Oil & Fat

Used cooking oil & fat cannot be composted in the standard 'Dalek'-type compost bin.

However, it can go into most other enclosed composting systems, such as the EM Bokashi system, tumbler compost bins, wormeries, digesters and Green Johanna-type bins; just mix with absorbent cardboard/paper before adding. You can compost it in a compost heap, providing you don't add too much at a time. Just make a hole in your heap and pour it in, covering it over with vegetable matter afterwards. For further details see under *Composting Systems*, page 22.

You can also thin down oil with some paraffin and use it to coat a wooden compost bin, to help preserve it.

Couch Grass

See Pernicious Weeds and Turf.

Diseased Plants

You need a well managed, regularly turned, hot system to tackle diseased plants (which is why most gardening experts will tell you to burn them), but the bacterial 'fire' of the compost heap is just as effective!

Composting is an amazing process: an astonishing number and variety of organisms are involved, and the heat generated can easily reach 60°–70°C. Plant and animal pathogens, as well as weeds, are destroyed by the heat generated by the heap, and the hotter it gets, the more quickly they die.

Weed seeds and human and plant pathogens are killed during the first few days of hot composting, when the temperature is above 55°C. But even just above 40°C, few will have survived after a month or so.

To ensure complete pathogen removal, it's important to mature your compost. The warm, post-hot composting phase produces the 'hygienisers', i.e. the organisms that attack pathogens.

Dog and Cat Faeces

These can contain parasitic worms that can cause blindness, and therefore it is generally not recommended that they are composted, especially when there are children around. However, a carefully made hot compost will destroy these pathogens; always cover them up in the heap or add to an 'in-vessel' system, which stops flies getting in.

If your council has a kerbside collection scheme for compostable materials, they are, quite understandably, not willing to take products such as these because of all the possible health risks. **Do not put them out for collection.**

Drinks Cans

These cannot be composted: recycle them at your local recycling centre.

Dryness

If you dig into your heap and find it dry and musty, don't dig any further; soak it with water first so that you don't breathe in lots of spores from fungi, then dig it out and re-stack, adding plenty of 'green' matter. Cover up with plastic or carpet to retain moisture.

Eggshells

The calcium in eggshells makes a useful additive to the heap, helping neutralise acid conditions in the heap and soil.

However, they will break down more rapidly if you crunch them up a bit as you put them in your kitchen caddy.

Eggshells are very popular with worms; the half shells make little refuges for them!

Fish

See Meat & Fish.

The Incas used to bury a fish under sweet corn seeds to help fertilise the growing plant.

Flats

If you live in a flat or apartment and do not have access to any garden or outside space, you can still use a wormery to compost your kitchen waste. You can also use a Bokashi system if you have somewhere to use the compost. See *Composting without Gardens* on page 49 for further information.

Flies

You may well find that you get masses of tiny flies that fly up when you lift the lid of your bin. These will mostly be harmless fruit flies. One way to control them is to leave the lid off the bin (or half off) for a while. This allows access to larger predatory beetles, which will set up home in your bin and will soon be feasting on fruit fly larvae. You can also wrap fruit waste in paper, or bury it in the heap – don't be tempted to use fly spray!

You usually get a few flies around compost heaps; if you get a lot then they are being attracted by something – larger flies are attracted by meat, manures and cooked kitchen scraps, which is why you should make sure you use the correct type of composting system if you wish to compost this type of food waste. See *Composting Systems* on page 22 for further information. If you are using a compost heap, cover it to deter flies.

Food

See specific food item, e.g. Meat & Fish, Vegetables etc.

Fruit

All fruit and its peel will compost, even citrus peel. Non-organic fruit is sprayed with preservatives, and citrus fruit is also treated with wax; both these factors inhibit the composting process.

If you make fruit juice or have enormous quantities of fruit peel and pulp, make sure it is well mixed with other materials, especially absorbent 'brown' materials like cardboard and paper, before adding to your compost system. It will be very acid, so add something alkaline – see Additives.

Fungi

Mushrooms and other fungi are fine in the compost heap. In fact, you may well see fruiting fungi of various types in your heap, and the white threads of their 'roots', called mycelium – this is quite natural. Your compost heap will contain hundreds of thousands of fungi, mostly microscopic, all doing a great job breaking down materials in your heap.

Garden Waste

All plant material will compost – don't listen to anyone who says that you can't compost rhubarb or ivy, for example. However, some plants – particularly pernicious weeds – need to be handled with extreme care, and others do have irritating sap or hairs – these also need treating with caution (see also Pernicious Weeds and Weeds).

Woody garden prunings and branches will take a very long time to break down, unless they are shredded or chipped up first. You can use a lawn mower on some of this kind of material, but obviously not on thick branches!

Large pieces of wood are often better used as firewood, or as wildlife refuges – many beetles depend on this kind of habitat. Some, like the stag's horn beetle, have become very rare because of people being over-zealous in clearing rotting lumps of wood from their gardens. Newts and toads also spend lots of time under rotting piles of wood, so find a place to make a wildlife pile, instead of trying to compost it all.

Also see Grass Cuttings, Leaf Mould and Shredders & Chippers.

Giant Hogweed

This non-native species is extremely pernicious and difficult to kill. **Do not compost or place out for kerbside collection.** You could be prosecuted for disposing of this weed incorrectly. See Pernicious Weeds for further advice and information.

Glass

This cannot be composted. If you do not have a kerbside collection, recycle it at your local recycling centre.

Grass Cuttings

Leave them on the lawn! Although grass cuttings are really useful in composting systems, they can also be a headache, especially in large quantities. You have to incorporate them into your heap in very thin layers (this is where having a stockpile of cardboard or woody chippings comes in handy, see page 20) and you need to do it as soon as they are cut. So why not just leave them on the lawn instead? As long as the grass is not too long it will not form a thatch on the top.

Continually removing grass cuttings impoverishes the soil, which may be great for creating wildflower meadows, as they thrive on impoverished soils, but lush lawns need fertile soil to thrive.

Hair

Do the hairdressers a favour and take your hair home to compost – all hair will compost, as it is rich in nitrogen, so all the pet hairs can go in too.

Hay

If you are lucky enough to be offered some hay bales, you can use them as building blocks for a super-insulated compost bin. Or you can break them apart and use them as a mulch around plants, or incorporate them into your compost heap. As with anything in quantity, mix well with plenty of other 'green' materials and make sure there is enough water.

Himalayan Balsam

This non-native species is extremely pernicious and difficult to kill. **Do not compost or place out for kerbside collection.** You could be prosecuted for disposing of this weed incorrectly. See Pernicious Weeds for further advice and information.

Insects

The compost heap is a natural magnet for all kinds of wildlife including insects, so don't worry unless you get an infestation (see under individual headings).

Japanese Knotweed

This non-native species is extremely pernicious and difficult to kill. **Do not compost or place out for kerbside collection.** You could be prosecuted for disposing of this weed incorrectly. See Pernicious Weeds for further advice and information.

Kitchen Waste

See specific food item e.g. Meat & Fish, Vegetables, etc.

When you are preparing vegetables for cooking, all the tops and tails and peelings can go in the ordinary 'Dalek'-type compost bin. However, although cooked food, meat and fish, cheese fats and grease can be composted, they need to go into one of the enclosed composters or fermenters as they can attract unwanted visitors. See *Composting Systems*, pages 22–39.

Kitchen Caddies

There are now several types of kitchen caddy bins. Some have activated charcoal filters to help control smells; others are designed to be used with compostable starch bags and have perforated walls to allow for airflow. Unperforated bins can be lined with several thicknesses of paper (newspaper is fine), which will help absorb any liquids and keep the bin clean. However, the easiest way to avoid odours is to empty the caddy frequently into your compost bin.

Leaves and Leaf Mould

Making leaf mould is very simple: you just gather up leaves and put them somewhere to stop them blowing around! In a large garden, the common solution is to build a wire enclosure to be filled up with leaves. The only additions you need are water and time. The leaves are broken down gradually – some take several years to fully break down, so you need to keep collecting them each year.

If you only have a small garden, you may prefer to fill plastic bags with leaves instead. Make sure the leaves aren't bone dry – there must be enough water for them to decompose.

See also *Leaf Mould*, pages 34–35.

Meat & Fish

Meat and fish scraps cannot be added to the normal 'Dalek'-type compost bin, as they can smell and attract flies and other larger unwanted visitors. These materials need to be composted within an enclosed composting or fermenting system. See *Composting Systems* on pages 22–39.

Manure

Manures from herbivore animals are full of nitrogen and are excellent for turbo-boosting your compost.

Manure is still usually wet, heavy and dense and is generally available mixed with straw, wood shavings or shredded paper/cardboard. It needs the same kind of treatment as grass cuttings: i.e. mix or layer well with anything that will help introduce air, absorb moisture and add carbon.

Poultry manure is strong stuff, so layer or mix even more thinly than horse, cow, donkey manure etc.

Faeces from meat-eating animals such as cats and dogs need special treatment, as they can contain parasitic worms and other pathogens (see Dog & Cat Faeces). However, 'Zoo Poo' from Paignton Zoo in Devon contains lion and tiger poo, and is supposed to keep cats off your flower beds!

Metals

These cannot be composted; if you do not have a kerbside collection, recycle them at your local recycling centre.

Mice

A compost heap is a cosy hotel for a mouse family, often providing free food and a roof! So don't encourage them by putting food in a heap that is easily accessible to them.

Monbretia

Also see Pernicious Weeds. Monbretia is a very difficult weed to kill off. Put it in heavy-duty plastic sacks and exclude light. Leave for about two years, and then it can safely be composted.

Nettles

Nettles are not difficult to find, and make a really useful addition to your bin or heap. They will add bulk, moisture, nitrogen, and plenty of beneficial minerals too. Their roots, however, need special care: see Pernicious Weeds and Activators.

Nappies

- If your council has a kerbside collection scheme for compostable materials, they are, quite understandably, not willing to take products such as these because of all the possible serious health risks related to dealing with blood and other body fluids. **Do not put them out for collection.**

- The best option is to use reusable washable nappies; either buy and wash them yourself, or use a nappy laundry service.

- But if you are a keen composter, you can now buy nappies made with bio-degradable, fully compostable plastic. It must say 'compostable' – not just 'degradable'. These can then be composted; but **NOT** in a normal 'Dalek'-type bin. Use a totally enclosed system, such as a Green Johanna. See also *Composting for the Serious Enthusiast*, especially *Hot Heaps* on pages 36–39.

- Urine is safe enough, but faeces need to be composted carefully, with plenty of other material to make a suitable mix.

- If you don't want to compost the faeces, soiled liner contents can be flushed down the toilet, and the rest of the nappy composted.

Nearly all disposable nappies are an environmental time-bomb, as well as being unpleasant and hazardous items in the dustbin. Buried in landfill sites they are said to take centuries to break down. Don't attempt to compost these disposable (more accurately termed 'one use') nappies.

Nitrogen

See Carbon to Nitrogen Ratio.

Oil

Vegetable oils used in cooking can be composted – see Cooking Oil – **but** new or used engine oil cannot be composted. Recycle it at your local recycling centre.

Paper

Any scrunched-up paper can be composted. Most composting experts tell you not to put highly glossy magazine paper in (it's best recycled), but the odd bit is not going to hurt. Most inks used these days are benign, and the glossiness is mostly fine clay particles. Most paper contains clay, which makes a smoother surface, and clay is fine for compost heaps. (Although clay soils are hard to work, they are the most fertile).

Pernicious weeds

(Also see under named weeds, e.g. Japanese Knotweed, and see Weeds for general advice)

There are a number of extremely pernicious weeds that require special treatment, as they are very difficult to kill and spread very easily. Indeed, it can be an offence to even grow some of these! They are Ragwort, Spear Thistle, Curled Dock, Broad-leaved Dock, Giant Hogweed, Japanese Knotweed and Himalayan Balsam. **Do not put them out for collection.**

It is illegal to take the last three weeds in this list anywhere, even to a centralised composting site. All the weeds listed above, therefore, must not be composted, either at home, or via your kerbside collection for compostable materials, if you have one.

These weeds must be destroyed, preferably by burning in your garden (if this is allowed where you live), or by sealing in plastic bags and taking to a licensed waste disposal site. You must tell the site operator what it is you have brought for disposal, and it will then be disposed of appropriately. Do not place in your domestic household waste bin for landfill. For further information, see www.environment-agency.gov.uk.

But for all other pernicious weeds, unless you are confident that you have a really hot heap which will cook pernicious weeds, it's probably best to exclude them from your compost heap.

Plastic ⊗

Avoid it – it will not compost, even if it says 'degradable'.

However there are some fully biodegradable, compostable plastics. These are commonly made from potato and/or maize starch, and are used to line kitchen caddies amongst other things. But unless you are 100% sure that your bag is compostable, leave it out of the compost heap.

Raw Food

Raw fruit and vegetables can be composted in a normal 'Dalek'-type or tumbler compost bin. All other raw food needs to be composted within an enclosed composting or fermenting system. See *Composting Systems*, pages 22–39.

Sanitary Towels

- If your council has a kerbside collection scheme for compostable materials they are, quite understandably, not willing to take products such as these because of all the possible serious health risks related to dealing with blood and other body fluids. **Do not put them out for collection.**

- If you are a keen composter you can compost them yourself in your garden. You can now buy sanitary towels that are bio-degradable and fully compostable. It must say 'compostable' – not just 'degradable'. These can then be composted, but **NOT** in a normal 'Dalek'-type bin. Use a totally enclosed system such as a Green Johanna. See also *Hot Heaps* pages in *Composting for the Serious Enthusiast* on pages 36–39.

Sauces

See Cooked Food.

Sawdust

Sawdust is very high in carbon (i.e. a 'brown' material), so it needs a lot of nitrogen – 'green' materials – to get it activated (see Carbon to Nitrogen Ratio). It takes a lot of wetting – you can add it very sparingly to compost heaps.

Seaweed

It is worth collecting seaweed, as it is a wonderful additive for any heap, with an amazing complement of nutrients and trace elements. The best time to collect fresh weed is immediately after a storm, or directly from the sea itself. Don't collect seaweed from the beach at other times, as it is likely to have a high salt content from salt spray building up and becoming concentrated. Salt is toxic to soils and most plants, and it is difficult to wash the salt off without contaminating the soil and watercourses.

If you can't get any from the beach, it is worth buying a bag of seaweed meal and sprinkling some in your heap. Or you can buy seaweed extract; water it down to feed plants (as with comfrey juice), and put it on your heap as well.

Shredders & Chippers

Shredders crush or hammer materials, whereas chippers cut them up. They enable you to reduce heavy duty prunings and other materials to a fraction of their bulk, and create a material with a massively increased surface area for composting. They range from the small, electric, almost silent, crushing type shredders, to very noisy petrol- or diesel-powered chippers.

Shellfish

See Fish.

Shells (from shellfish)

The calcium carbonate from the shells of shellfish will help neutralize acidic soil, prevent blossom end rot (a calcium deficiency), and is said to yield a better-tasting tomato.

If you have a shredder, it may be powerful enough to smash up shells from mussels, oysters, crabs etc, or you can get your hammer out. Otherwise, bury them in your composter or wormery – the harder shells will take a very long time to break down, and will keep turning up in your garden, but will eventually disappear.

Shells (from eggs)

See Eggshells.

Slugs & snails

You get some quite spectacular-looking slugs attracted to compost heaps. Most slugs feast on wilted or decaying vegetation. The ones that go for living plants are small, orange and black or greyish in colour. You can just leave them alone on your compost heap, as they help to break down materials and are positively beneficial, as are snails.

Smell

Compost heaps shouldn't smell unpleasant. If they do, then they need more 'brown', carbon-rich materials mixed in. You may have to dig out or turn the heap and mix in something dry and absorbent to open it up: cardboard, sawdust, wood chippings, dry weeds, paper etc.

Soot

See Wood Ash.

Starch Bags

Bags made from maize or potato starch are totally biodegradable and therefore compostable. Both kitchen and other green waste can be collected within these bags, allowing both bag and contents to be placed in the composting system or heap.

Stirrer

You may be given a device that looks like a harpoon with your compost bin. You plunge this into your compost, and it aerates the compost as you pull it out, which is much easier than digging the whole heap out.

Straw

You can use straw bales to make the sides of a wonderfully insulated compost heap. Alternatively, you can compost the straw by pulling the bales apart and using them as part of your 'browns' mix to add carbon to your pile. See *What do I put in my bin?* on page 15, and *How do I fill it?* on page 16. Straw can also be used as a mulch around shrubs, fruit bushes etc.

Tea Bags

Not all tea bag material will compost, so it's best to break them open (easiest when dry), shake out the tea and discard the bag – or use loose tea!

Tins ⊗

These cannot be composted. If you do not have a kerbside collection, recycle them at your local recycling centre.

Tissues

If your council has a kerbside collection scheme for compostable materials, they are, quite understandably, not willing to take products such as these because of all the possible health risks related to dealing with body fluids. **Do not put them out for collection.**

- Keen composters can compost used tissues; but **NOT** in a standard Dalek-type bin. Use a totally enclosed system such as a Green Johanna. See also *Composting for the Serious Enthusiast*, especially *Hot Heaps*, pages 36–39.

Turf

You can make a loam (good quality garden soil) with turf – even from weed-infested turf like couch grass. Stack it neatly upside down, cover with heavy-duty black plastic and leave for a year or two. It will turn into wonderful crumbly loam. Sieve before use, and use as ordinary garden soil. Another option is to make turf chairs or a bench – just use the turfs like bricks.

Vacuum Dust

Providing you do not have synthetic carpets (which will not compost), the contents of your vacuum cleaner can be easily composted by any compost system. See also Carpets.

Vegetables

Raw and cooked – see Kitchen Waste.

Water

The main trick with composting is to get the mix right at the start. This means balancing the 'greens' with the 'browns' – balancing the wet and the dry materials. If this is roughly right, the environment will be right for all the organisms that drive the compost system.

- If the heap is too dry, many of the organisms just die or go away, and the compost stops working. If this happens you will need to add some water.

- If the heap is too wet (usually the problem), just make sure you add plenty of absorbent paper and cardboard. See *Now What?* on page 16.

Weeds

See also Pernicious Weeds.

Weeds are great to add to compost heaps as long as they are not seeding profusely, and you are very careful what you do with the perennial pernicious ones. Indeed, some require special treatment and it is illegal even to move them elsewhere. Weeds tend to be wet and sappy, so try and mix them with some 'brown' material, and always remove as much soil as you can from them, as this helps them to compost more readily and you don't want too much soil in a compost heap.

Weeds are also valuable in the compost, as many are 'dynamic accumulators', which means that they have the ability to concentrate minerals that are in low concentrations in the soil, thus making them more readily available for other plants. Some of them do this by having very deep roots, which bring up the minerals that are out of the reach of other plants, so don't burn them – just treat with respect. Weeds are hardy, and can go to seed even after being pulled out the ground, or re-root if left on a damp soil. See also Pernicious Weeds for further details about dealing with weed roots.

However, weed roots contain lots of valuable minerals, so it's a great shame to waste the roots of non-pernicious weeds:

- Dry them out in the sun for a few days before composting, but ensure the attached plant doesn't go to seed.

- For very earthy weed clumps like couch grass, make a neat stack of the earthy clods, and then cover up the whole heap with thick black plastic to exclude all light. Leave for about two years to be sure that all the weeds have died, and the result will be a lovely rich loam. Smaller amounts can be treated in light-proof bags.

- Drown the weeds to release their minerals. They can either be put in a hessian sack and weighed down in a barrel of water, or simply put into a barrel of water with a cover on it. After a few weeks, the whole lot will rot and smell awful – so do it well away from human habitation! The smelly water can be used on plants (the smell will go away very quickly), and the plant remains can now be safely composted.

Weed Facts

- Plants die without light

- Most plants die without water, so dry them out on a wire frame – beware of bulbs and corms, though, such as Monbretia, which is difficult to kill off.

- Most plants will 'drown' in water.

- Fly tipping is illegal, and causes the spread of invasive weeds.

Wetness

Compost systems need water, but they mustn't become waterlogged. The water must not fill up all the airspaces – it must be either absorbed (preferably), or allowed to drain away, as otherwise everything drowns and it really starts to smell! If this happens you will have to dig it all out and mix with absorbent and free-draining high carbon materials, like cardboard and wood chippings. This is why it is vital to drain totally enclosed systems such as wormeries on a regular basis. See also Anaerobic.

Wood Ashes & Soot from Wood Fires

These are fine to add to a compost heap a little at a time – they add valuable potash, although adding too much at any one time will result in a nasty sludge.

Worms

Worms should invade a healthy heap. There are several sorts:

- Small white thread worms, which you will also find in your wormery

- 'Tiger worms', which are red with yellow stripes

- Other types of red manure worms

- The large mauve garden earthworm, although this only appears occasionally – it is usually in the soil.

Worms will live in cool moist conditions, and will burrow away from any hot composting that may be going on.

10

Resources

BOOKS & VIDEOS

Composting for All Video with Nicky Scott, Green Books.

Liquid Gold: The lore and logic of using urine to grow plants by Carol Steinfeld, Green Books.

A Guide to Community Composting published by the Community Composting Network – see below.

ORGANISATIONS

The Community Composting Network

The Community Composting Network supports and promotes the community management and use of waste biodegradable resources. Projects range in scale from individuals or small groups working on allotment sites or promoting home composting, to social enterprises with Local Authority contracts providing kerbside collection services.

The Community Composting Network,
67 Alexandra Road, Sheffield S2 3EE
Tel 0114 2580 483 www.communitycompost.org

The Composting Association

The Composting Association is the United Kingdom's not-for-profit membership organisation, promoting the sustainable management of biodegradable resources. It actively promotes the use of biological treatment techniques, and encourages good management practices throughout the industry.

The Composting Association,
Avon House, Tithe Barn Road, Wellingborough,
Northants NN8 1DH
Tel: 0870 160 3270 www.compost.org.uk

Biodynamic Agricultural Association (BDAA)

The BDAA is founded on a holistic and spiritual understanding of nature and the human being's role within it.

At its heart is the idea of the farm as a self-contained evolving organism, whose life relies on home-produced compost manures and animal feeds, with minimum external inputs.

Biodynamic Agricultural Association,
Painswick Inn, Stroud, Glos GL5 1QG
Tel: 01453 759501 www.anth.org.uk/biodynamic

Defra

Defra (the Department for Environment, Food and Rural Affairs) is a government department that deals with food, air, land, water, people, animals and plants. Contact them or see their web site for advice, including the safe disposal of pernicious weeds.

Defra Information Resource Centre,
Lower Ground Floor, Ergon House,
c/o Nobel House, 17 Smith Square,
London SW1P 3JR
Tel: 08459 33 55 77 www.defra.gov.uk

Environment Agency

The EA is responsible for protecting and improving the environment in England and Wales. Contact them or see their excellent website for advice on the safe disposal of non-compostable materials, including pernicious weeds.
Phone 08708 506506 to find the address of your regional centre. www.environment-agency.gov.uk.

HDRA – The Organic Organisation

HDRA is dedicated to researching and promoting organic gardening, farming and food.

HDRA (Henry Doubleday Research Association),
Ryton-on-Dunsmore, Coventry CV8 3LG
Tel: 02476 303517 www.hdra.org.uk

Waste and Resources Action Programme (WRAP)

The WRAP Home Composting Scheme works with local authorities and other selected organisations to promote home composting. They encourage people to purchase compost bins at a subsidized price. Advice and support is also provided for all households engaging in home composting, via support materials, a dedicated helpline and advisors in the field.

Waste and Resources Action Programme
The Old Academy, 21 Horse Fair, Banbury, Oxon OX16 0AH
Home Composting Helpline: 0845 600 0323
www.wrap.org.uk